A VISITORS' GUIDE TO

KIRSTENBOSCH

Text and photographs by Colin Paterson-Jones

Situated on the slopes of Cape Town's Table Mountain, Kirstenbosch is one of the world's great botanical gardens. It contains a unique collection of southern African plants and provides a gateway to the hikes along and up the mountain's eastern face.

This guide will help you to make the most of your visit.
If you have less than half a day, turn the page;
if you have longer, turn to pages 4/5.
For general information, turn to pages 31/32.

IN AN EMERGENCY:

Contact the information office (see the map overleaf) or one of the security personnel who patrol the garden.

National Botanical Institute
South Africa

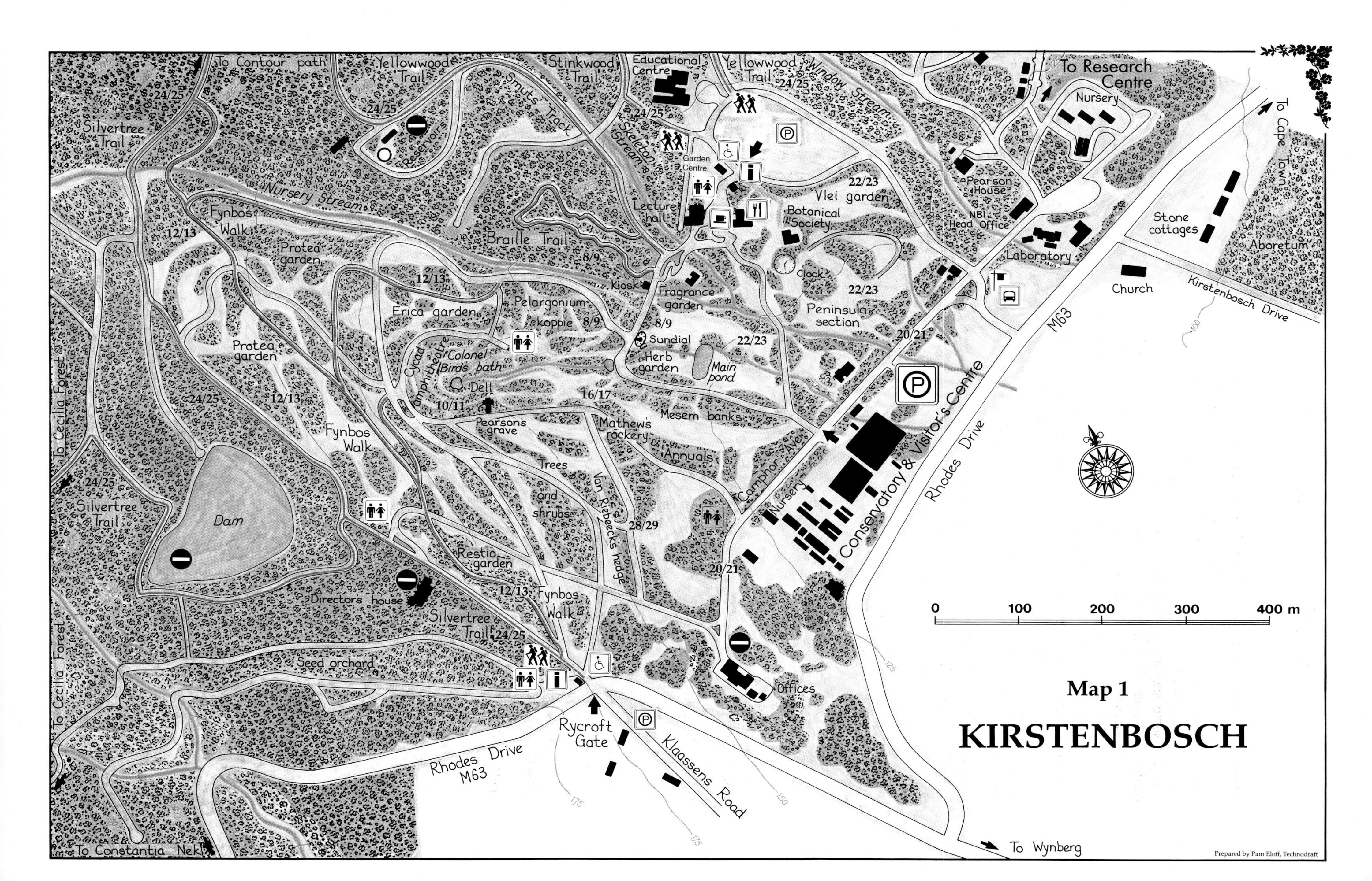
Map 1
KIRSTENBOSCH
0
100
200
300
400 m
To Cape Town
To Research Centre
Nursery
Stone cottages
Aboretum
Kirstenbosch Drive
Church
Laboratory
NBI Head Office
Pearson House
M63
Rhodes Drive
Conservatory & Visitor's Centre
Nursery
Camphor Ave
Offices
To Wynberg
Klaassens Road
Rycroft Gate
Rhodes Drive
M63
To Constantia Nek
To Cecilia Forest
Silvertree Trail
To Contour path
Yellowwood Trail
Stinkwood Trail
Smuts' Track
Reservoirs
Educational Centre
Yellowwood Trail
Window Stream
Skeleton Stream
Garden Centre
Lecture hall
Vlei garden
Botanical Society
Clock
Peninsula section
Nursery Stream
Fynbos Walk
Protea garden
Braille Trail
Kiosk
Fragrance garden
Sundial
Herb garden
Main pond
Erica garden
Pelargonium koppie
Cycad amphitheatre
Colonel Bird's bath
Dell
Pearson's grave
Mathew's rockery
Mesem banks
Annuals
Trees and shrubs
Van Riebeeck's hedge
Restio garden
Dam
Directors house
Silvertree Trail
Seed orchard
Fynbos Walk
24/25
12/13
8/9
22/23
20/21
16/17
10/11
28/29
Prepared by Pam Eloff, Technodraft

VISITS OF ONE TO THREE HOURS

Kirstenbosch can be entered either through the Rycroft Gate, the Garden Centre gate and upper car park, or through the Visitors' Centre. All parts of the Garden have a wealth of interest at any time of the year but some sections are especially colourful in particular seasons (indicated in brackets). No visit should leave out the Botanical Society Conservatory [33] which alone has enough of interest to occupy a rainy day. To appreciate the Garden fully requires days, even weeks, throughout the year. If your time is limited, however, we suggest:

Note: Numbers in square brackets refer to points on the map as well as page numbers in this guide.

ONE-HOUR VISITS FROM THE RYCROFT GATE TO THE VISITORS' CENTRE CAR PARK (MOSTLY DOWN-HILL OR LEVEL WALKING)

1 Take the Fynbos Walk through the restio garden, protea garden (winter, spring, early summer) and the erica garden [12/13], then visit the cycad amphitheatre and Dell [10/11] and the pelargonium koppie (spring) [8/9]; or

2 Take the path to the cycad amphitheatre and Dell [10/11], then visit the erica garden [12/13], pelargonium koppie (spring) [8/9], Mathews' rockery (winter, spring) and the mesem banks (October) [16/17], Camphor Avenue (summer, autumn) [20/21] and Peninsula section [22/23]; or

3 Take the path past Van Riebeeck's Hedge [28/29] to Mathews' rockery (winter, spring) and the mesem banks (October) [16/17], Camphor Avenue (summer, autumn) [20/21], main pond and lawns, Peninsula section and the vlei garden (spring, early summer) [22/23].

ONE-HOUR VISITS FROM THE VISITORS' CENTRE OR THE GARDEN CENTRE AND BACK (LEVEL, UPHILL AND DOWNHILL WALKING)

1 Visit the pelargonium koppie (spring) [8/9], erica and protea gardens (winter, spring, early summer) [12/13], cycad amphitheatre and Dell [10/11]; or

2 Visit the pelargonium koppie (spring) [8/9], cycad amphitheatre and Dell [10/11], Mathews' rockery (winter, spring) and the mesem banks (October) [16/17], Camphor Avenue (summer, autumn) [20/21], main pond [22/23] and the herb garden [8/9]; or

3 Visit the fragrance and herb gardens [8/9], Mathews' rockery (winter, spring) and the mesem banks (October) [16/17], Camphor Avenue (summer, autumn) [20/21], main pond, Peninsula section and the vlei garden (spring, early summer [22/23].

ONE-HOUR VISIT FROM THE GARDEN CENTRE

Walk the Stinkwood Trail [24/25].

THREE-HOUR VISITS FROM THE RYCROFT GATE TO THE VISITORS' CENTRE OR THE GARDEN CENTRE (MOSTLY DOWN-HILL OR LEVEL WALKING)

1 Visit the fynbos area including the restio garden, protea garden (winter, spring, early summer) and erica garden [12/13], pelargonium koppie (spring) [8/9], cycad amphitheatre and Dell [10/11], Van Riebeeck's Hedge [28/29], Camphor Avenue (summer, autumn) [20/21], the mesem banks (October) and Mathews' rockery (winter, spring) [16/17], and the herb and fragrance gardens [8/9]; or

2 Visit the cycad amphitheatre and Dell [10/11], erica garden [12/13], pelargonium koppie (spring) [8/9], Mathews' rockery (winter, spring) and the mesem banks (October) [16/17], Camphor Avenue (summer, autumn) [20/21], the main pool and lawns, Peninsula section and the vlei garden (spring, early summer) [22/23].

THREE-HOUR VISIT FROM THE RYCROFT GATE AND BACK (UPHILL, LEVEL AND DOWNHILL WALKING)

Walk the Silver Tree Trail [24/25].

THREE-HOUR VISITS FROM ANY ENTRANCE AND BACK (UPHILL, LEVEL AND DOWNHILL WALKING)

1 Visit the fragrance and herb gardens and the pelargonium koppie (spring) [8/9], the cycad amphitheatre and Dell [10/11], the erica and protea gardens (winter, spring, early summer) [12/13], Van Riebeeck's Hedge [28/29], Camphor Avenue (summer, autumn) [20/21], the mesem banks (October) and Mathews' rockery (winter, spring) [16/17]; or

2 Visit the fragrance and herb gardens and the pelargonium koppie (spring) [8/9], the erica garden [12/13], the cycad amphitheatre and Dell [10/11], Mathews' rockery (winter, spring) and the mesem banks (October) [16/17], Camphor Avenue (summer, autumn) [20/21], the main pond and lawns, Peninsula section and the vlei garden (spring, early summer) [22/23].

THREE-HOUR VISIT FROM THE GARDEN CENTRE

Walk the Yellowwood Trail [24/25] and visit the fragrance and herb gardens and the pelargonium koppie (spring) [8/9], and the cycad amphitheatre and Dell [10/11].

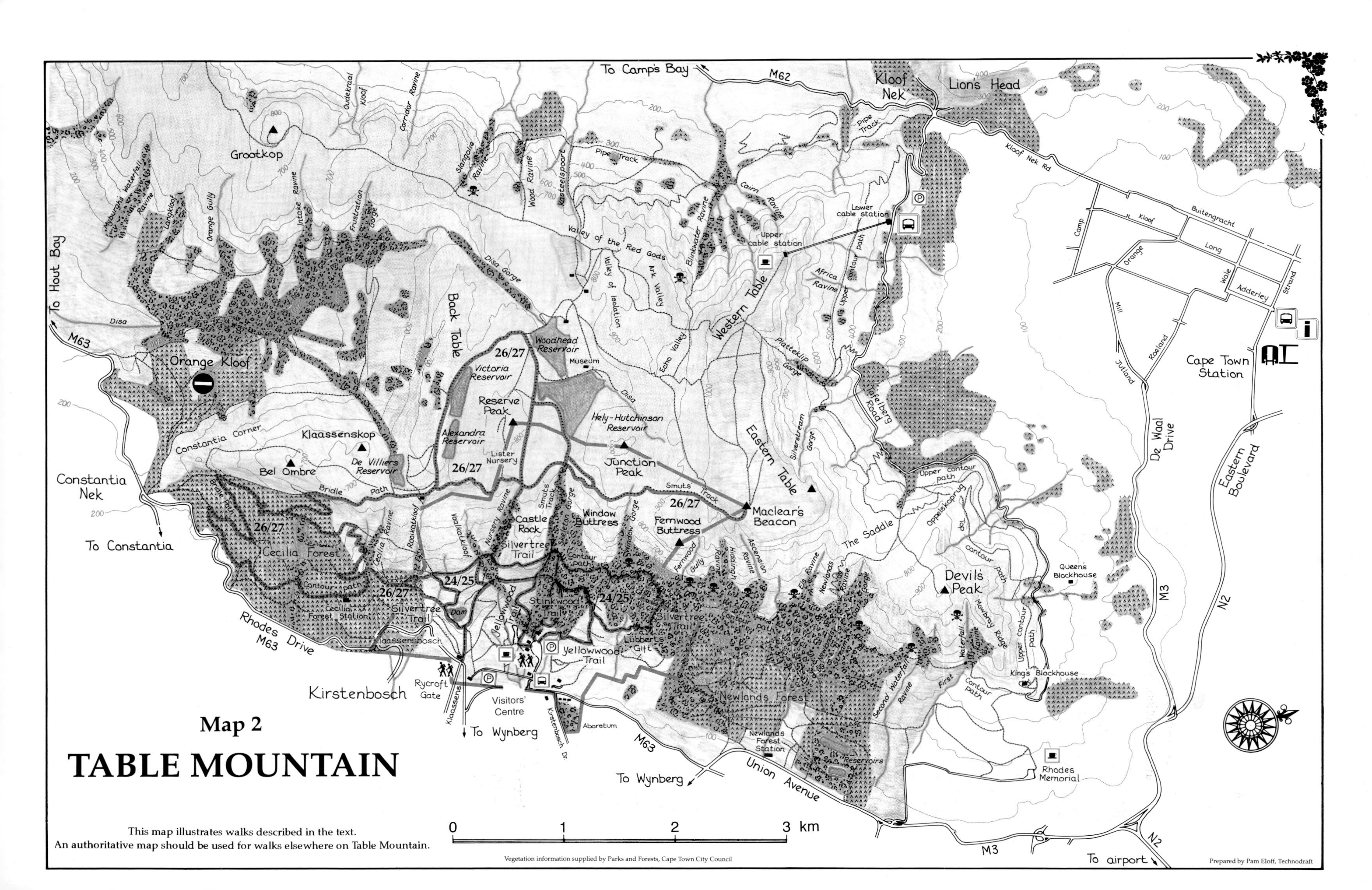

Map 2
TABLE MOUNTAIN

This map illustrates walks described in the text.
An authoritative map should be used for walks elsewhere on Table Mountain.

Vegetation information supplied by Parks and Forests, Cape Town City Council

Prepared by Pam Eloff, Technodraft

HALF- AND FULL-DAY VISITS

HALF-DAY (5-HOUR) VISITS

1 In five hours you can see all of the cultivated area of Kirstenbosch. Pages 8 to 23 provide information on the different sections of the Garden; or

2 From the Rycroft Gate hike via Cecilia Ravine into Cecilia Forest and back along the contour path [26, 27]; or

3 From the Rycroft Gate or the information office, hike up Nursery Ravine to the Woodhead Reservoir and return via the Bridle Path, Cecilia Forest and the contour path [26/27]; or

4 From the Rycroft Gate or information office, hike up Skeleton Gorge to the Hely-Hutchinson Reservoir and return via Nursery Ravine [26/27].

FULL-DAY (8-HOUR) VISITS

1 A full day provides ample time for a leisurely walk round the cultivated section of Kirstenbosch (see pages 8 to 23) as well as one of the forest walks [24/25]; or

2 Take Smuts' Track, climb up Skeleton Gorge and up to Maclear's Beacon and return via Nursery Ravine [26/27].

SAFETY ON THE MOUNTAIN

Every year climbers and hikers are killed or injured on Table Mountain. These accidents can be avoided if hikers are properly equipped and dressed, and take some elementary precautions. The eastern flank of Table Mountain is precipitous with a number of very steep gorges and sheer faces which are dangerous to any but experienced and suitably equipped rock climbers. Only two gorges provide a safe route up and down this part of the mountain – Nursery Ravine and Skeleton Gorge – while access in the south-east is provided by the Bridle Path.

☐ Do not walk alone; three people is a safe minimum group; keep together.
☐ Tell someone where you are going and stick to this route; keep to well-known and unobstructed paths.
☐ Use a reliable, accurate map.
☐ Always carry warm, windproof clothing – cloud can come up in minutes on an apparently clear day.
☐ Wear sensible clothes and sturdy shoes or boots.
☐ Always carry water.
☐ If you become lost, particularly in the mist or dark, *do not panic* – find shelter, keep warm and dry and wait until you can see where you are.
☐ Do not exceed the physical capabilities of the weakest member of your group.

Please note that littering, writing graffiti, playing loud radios or tapes, lighting fires, discarding cigarette ends, and picking flowers are both unacceptable behaviour and illegal.

WEATHER

Kirstenbosch's weather is temperate. Rain, brought in by cold fronts, falls mainly in winter and spring (annual average rainfall 1 450 mm, average number of rainy days per year 140). Winters are mild (average daily minimum 8,5 °C, average daily maximum 16,8 °C in August, the coldest month), with occasional strong north- and south-westerly winds. Summers are moderate (average daily minimum 15,4 °C, average daily maximum 24,7 °C in February, the hottest month). The absolute minimum temperature recorded is 1,4 °C, the absolute maximum 38,6 °C. Strong, desiccating south-easterly winds can persist for days in summer but the Garden is sheltered from their full force.

The tall-growing *Psoralea pinnata* cohabits with *Berzelia lanuginosa* and *Erica curviflora* in fynbos along Smuts' Track.

THE CAPE FLORA

The plants of the south-western and southern Cape (between the Great Karoo and the sea) are remarkably different from the plants found further north in southern Africa, Africa as a whole and, in fact, the world. The uniqueness of this assemblage of plants – known as the Cape Flora – which is found in an area of only 90 000 km^2 (less than 4 per cent of the area of southern Africa), has led to its being considered a floristic kingdom, ranked in importance with the five other floral kingdoms of the world – the Holarctic, Palaeotropic, Neotropical, Australasian and Antarctic kingdoms – each of which covers a vastly greater area of the earth's surface.

There are about 8 500 species of plants in the Cape Flora (some 42,5 per cent of the total for the whole of southern Africa) and, of these, about 68 per cent are found nowhere else. Six plant families – the Penaeaceae, Grubbiaceae, Stilbaceae, Roridulaceae, Geissolomaceae and the Retziaceae – are unique to the Cape Flora and one family, the Bruniaceae, has only one species elsewhere.

The Cape Flora includes several vegetation types. A

unique type is fynbos, a vegetation adapted to poor soils and periodic fires, which is characterised by the presence of hard- or small-leaved shrubs including a large number of species from the protea family (Proteaceae), erica family (Ericaceae) and restios (Restionaceae). Restios take the place of grasses and sedges which elsewhere in southern Africa are the dominant ground cover. There are also many species of the daisy (Asteraceae) and legume (Fabaceae) families. The other vegetation types are strandveld, a shrubby vegetation found on lime-rich soils; succulent karoo veld which contains numerous succulent plants such as the mesems (Mesembryanthemaceae); renosterveld, dominated by the renosterbos (*Elytropappus rhinocerotis*) and containing a wealth of bulbous plants; and Afromontane forest.

Long-term climatic changes in southern Africa contributed to the evolution of the Cape Flora. Since some 26 million years ago, southern Africa has become cooler and drier, with a seasonal climate. Until about 3 million years ago, the Cape was probably covered with evergreen forest similar to the present-day forests of southern Chile, New Zealand and Tasmania - the Knysna forests are a remnant. About 1,5 million years ago, the aridity reached a peak and a winter rainful climate became established in the south-western Cape. The ancient forests shrank and more drought-resistant plants spread to take their place. These included the ancestors of the dominant families of the Cape Flora. From this time on there were large climatic fluctuations associated with the periodic expansion and contraction of the polar ice-caps. These climatic changes together with the frequent occurrence of fires, the broken terrain and varied and nutrient-poor soil types were the environmental pressures to which the plants had to adapt. The result was an explosion of varied plant forms as the plants evolved and diversified in a profusion of different, localised habitats. The result is a uniquely beautiful collection of plants which has delighted and fascinated botanists and plant lovers for over three centuries. Many of these plants can be seen at Kirstenbosch.

The restio *Elegia capensis.*

Leucospermum oleifolium, a member of the Proteaceae.

Protea aristata, the Christmas protea.

Erica phylicifolia, an erica of the Cape Peninsula.

Schotia brachypetala (boerboon), below the pelargonium koppie, with Castle Rock as a background.

THE PELARGONIUM KOPPIE AND ENVIRONS

The pelargonium koppie ('koppie' is Afrikaans for small hill) is found in a relatively dry part of Kirstenbosch, and between its rocks grows a profusion of plants including representatives of some of the drier types of fynbos. Pelargoniums are, of course, well represented and visitors can see the parent species of many popular garden hybrids. Because of the damage caused by Cape molerats and porcupines, not many parts of Kirstenbosch are suitable for growing small bulbous plants, but the pelargonium koppie is an exception and, particularly in spring, a delightful range of the Cape's bulbs is in flower here. There are not many species of *Aloe* in fynbos but on the koppie there are some shrubs of *Aloe plicatilis*, a species growing only in fynbos.

Next to the main path, opposite the pelargonium koppie, is the herb garden containing plants whose stems, leaves, flowers or roots can either be used medicinally, for flavouring or for perfume. Growing together with the aromatic plants in this garden are two legumes, *Cyclopia genistoides* (Heuningtee) and *Aspalathus linearis* (Rooibos), both fynbos plants whose leaves are used to make herbal teas. Rooibos tea has become so popular that it is now a significant export crop in the dry fynbos area of the Cape.

The display kiosk nearby exhibits named examples of flowers which can currently be seen in the Garden. Diagonally opposite this is the fragrance garden, and adjoining it the braille trail, two fascinating features which are of particular interest to the elderly, and the visually impaired. Roughly 470 m long, the braille trail is a 40-minute leisurely stroll, which takes you through a wooded, marshy area, home to the Knysna warbler, a secretive bird more often heard than seen. A guide rope leads the way, and at 10 stopping points there are signs in large print and braille describing interesting plants along the route.

Plants with aromatic leaves or special textures have been set out in waist-high beds in the fragrance garden, to be touched and smelt conveniently. Plant labels here are also in large print and braille. Aromatic fynbos plants of the Rutaceae family such as *Agathosma* (Buchu) and *Coleonema* (Confetti Bush) species are well represented. There are also fragrant-leaved pelargoniums, like the rose-scented *Pelargonium capitatum*, and local mint plants, like *Mentha longifolia* subsp. *capensis*, as well as a variety of plants with textured foliage.

Romulea flava.

Aloe plicatilis.

Pelargonium cucullatum.

The fragrance garden.

The cycad amphitheatre.

Colonel Bird's bath.

THE DELL AND THE CYCAD AMPHITHEATRE

The Dell has a character all of its own and is probably the most popular feature of Kirstenbosch for visitors. Permanent springs provide the clear water which flows into Colonel Bird's bath and then out around stepping stones down through the Dell to the small pond at the corner of the main path. The Dell is shaded by large yellowwood trees (*Afrocarpus (Podocarpus) falcatus, Podocarpus latifolius* and *P. henkelii*). A fourth species (*P. elongatus*) stands next to the stone bridge over the stream below the Dell, making up the collection of all four southern African yellowwood species. Above Colonel Bird's bath itself, which is surrounded by tree ferns (*Cyathea dregei* and the exotic *Dicksonia antarctica*), grows a huge tree of *Ilex mitis*, the Cape holly. This is the only Cape species of the hollies which are widespread in the northern hemisphere and well loved in cultivation there. True to its type, the Cape holly produces red berries which are prized by fruit-eating birds, including the rameron pigeons that frequent this area. Between the rocks that line the Dell grow shade-loving plants, including a large number of *Streptocarpus* species whose mauve flowers light up the dark shade in summer and autumn. Towards the small pond a variety of shrubs line the streamlet. One of these is *Gardenia thunbergii* whose white flowers perfume the Dell on a still summer's day.

Surrounding the Dell is the cycad amphitheatre. This was the first plant collection to be established at Kirstenbosch and dates from the time of its first director, Professor Harold Pearson, who had a special professional interest in these plants and other gymnosperms (non-flowering seed-bearing plants). Cycads have been labelled 'living fossils' – they flourished on earth before flowering plants had evolved. The cycad amphitheatre contains a fine collection of the southern African species. It is fitting that Professor Pearson is buried at the southern corner of the amphitheatre.

Looking down towards the Dell.

An olive thrush on the cobbled floor of the Dell.

Streptocarpus candidus.

Fernwood Buttress towers over the erica garden.

THE FYNBOS AREA

The higher slopes of the cultivated part of Kirstenbosch comprise a substantial area of lawns, with beds largely devoted to fynbos plants. Through this area runs the Fynbos Walk, a brickpaved path with gentle gradients which takes you from the Rycroft Gate in a broad sweep to the pelargonium koppie. The fynbos area is true to name in an important sense since only a protracted visit will reveal the extraordinary diversity of the species planted here - a representative reflection of the richness of the fynbos. The area is broadly divided into the restio garden near the Rycroft Gate, the protea garden and the erica garden, a subdivision which reflects the three characteristic fynbos families (see pages 6/7). In spite of the names, the plantings are not exclusively confined to these families, but also contain a huge wealth of fynbos plants of other families. There are also several summer rainfall proteas in the protea garden, including one of the most beautiful of these, a small tree of *Protea rubropilosa*, standing alone in the centre of a lawn.

The horticultural potential of the restios has only recently been realised through successful research on their propagation. The attraction of these unique and beautiful fynbos plants is displayed in the restio garden next to the Rycroft Gate. Restio species are also an integral part of the plantings in the protea, erica and Peninsula gardens [22/23]. A prominent species here is *Elegia capensis*.

Most of the fynbos protea species such as *Protea repens* and *P. neriifolia* flower in autumn and winter, but there are some spectacular exceptions; one of these is the Christmas protea (*P. aristata*) whose flowering peak is in December. Spring is the most colourful time in the protea garden when plants of other genera in the protea family, such as *Leucadendron*, *Mimetes* and *Serruria*, and of other families such as the legumes (*Podalyria* species) and Rhamnaceae (*Phylica* species) put on a spectacular show. This is also the season when the uniquely beautiful silver trees flower. Towards summer, the pincushions (*Leucospermum* spp. in the protea family) flower, to the delight of visitors, sunbirds and sugarbirds alike. In the wild the orangebreasted sunbird and Cape sugarbird are restricted to fynbos, and both these birds are resident in the fynbos area of Kirstenbosch.

The erica garden contains a large number of the over 600 southern African *Erica* species which ensures that there is colour here throughout the year as different species flower in succession. Some ericas are striking because of the masses of flowers they produce, others like *Erica regia* are notable because of the beauty of their individual flowers. Like the protea garden, this section is a community planting with an abundance of companion fynbos plants. The views from here across the Cape Flats to the Stellenbosch, Paarl and Hottentots Holland mountains are especially beautiful.

Silver trees (*Leucadendron argenteum*) on the lawns in the protea garden.

FAUNA

BIRDS

Kirstenbosch is well known as a bird-spotting venue. The cultivated area, the fynbos, the forest and the mountain slopes provide different habitats that attract a wide variety of birds. Sunbirds and the Cape Sugarbird play a major role in the pollination of many of the proteaceous and bulbous plants in the Garden. A checklist of birds which may be seen at Kirstenbosch is available at the information office (see map on page 2).

MAMMALS

Few mammals (apart from *Homo sapiens*!) are likely to be seen in the cultivated area of Kirstenbosch. The large grey mongoose patrols the paths in the early morning and evening, while the introduced grey squirrel is common but confined to the oaks. Striped mice are plentiful. Evident only from their respective underground runs and soil heaps on the lawns are Cape golden moles and Cape molerats. Above the cultivated gardens on the mountainside, the shy rock dassies (the favourite prey of black eagles) like to bask on the rocks. Baboons are sometimes encountered on the mountain.

REPTILES, AMPHIBIANS AND FISH

There are venomous snakes at Kirstenbosch but these are seldom seen; the most common is the boomslang, a shy retiring creature of the scrub and forest. Very occasionally, a Cape cobra, puff adder or berg adder is spotted. Non-venomous snakes include the common slugeater, olive house-snake, mole snake, and the common brown water snake, the last associated with the streams. Lizards are plentiful, the most obvious being the dark-coloured Cape girdled lizard and the southern rock agama; in the breeding season the male agamas are brilliantly coloured in blue, green and orange. Angulate tortoises are sometimes spotted. The much larger and occasionally seen leopard tortoises are not natural here but were introduced from the Little Karoo. The Cape river frog abounds in the streams and pools of the Garden. Higher up, in wet areas on Table Mountain, the beautifully marked banded stream frog is found. Table Mountain's endemic frog species, the thumbed ghost frog, only occurs in the perennial streams of its eastern flank. This frog is nocturnal and seldom seen. The Cape chirping frog is found in the vlei and along the braille trail. Cape platannas may be seen in the ponds in the Garden, which are also home to three species of fish – *Galaxias zebroides, Sandelia capensis* (Cape bream) and *Tilapia sparrmanni.*

INSECTS

The insect life of Kirstenbosch, while largely unnoticed by visitors, is prolific and important. Many of the flowers are insect-pollinated, and since seed is collected for propagation and distribution of the excess to Botanical Society members, the insects' activity is vital. In spring, the work of many kinds of bees is obvious, but it takes a keen eye to see some of the strikingly coloured scarab (cetoniid) beetles on the proteas and pincushions (*Leucospermum* spp.), and the blister beetles (*Mylabris* spp.) on a wide variety of other flowers. Butterflies abound in spring and summer. *Aeropetes tulbaghia*, the Table Mountain beauty, is the only pollinator of the red disa (*Disa uniflora*). In November, Kirstenbosch staff take visitors to see the fireflies in the forest after dark, if conditions are suitable.

Spotted dikkop.

Cape francolin.

Southern rock agama.

Helmeted guineafowl.

Male orangebreasted sunbird on *Leonotis ocymifolia*.

Spotted eagle owl.

Cape robin.

Protea scarab beetle.

Large grey mongoose.

Mesems provide a riot of colour in October.

A *Bulbine* species thrown into relief by a red vygie.

The aloes flower in winter on Mathews' Rockery.

The main path leads past the mesem banks, below Mathews' Rockery.

MATHEWS' ROCKERY

The first curator of Kirstenbosch, J W Mathews, was the man responsible for the building and planting of the rockery which bears his name. Its construction from Table Mountain sandstone boulders started in 1927. This area is devoted to succulent plants from the semi-arid and arid regions of southern Africa that can adapt to the relatively high rainfall of Kirstenbosch. The secret is drainage and the successful cultivation of a rich representation of aloes, euphorbias, crassulas, mesembryanthemums, and many other dry-area plants is a living testament to the skill of the rockery's construction. In addition to succulents, bulbous plants such as *Eucomis* species can be seen growing here. The area is striking in spring but is at its most colourful in winter, when the aloes flower and the lesser double-collared sunbird feasts on their nectar. For two or three weeks in October, the rockery and adjacent mesem banks are alight with the massed brilliance of mesembryanthemum (vygie) flowers – an unforgettable sight. Note that these flowers do not open properly, if at all, on overcast days.

Spring daisies next to the main lawn.

Gazania krebsiana.

Blister beetle on *Watsonia borbonica* subsp. *borbonica.*

Sparaxis sp.

Ursinia sp. surrounded by wine cups (*Geissorhiza radians*).

Spring daisies.

Agapanthus praecox subsp. *orientalis.*

SPRING FLOWERS

Massed displays of spring flowers are among the glories of Kirstenbosch. From July to September large areas of the Garden are ablaze with colour. Most of the early season colour in Kirstenbosch comes from plantings of the annuals of the daisy family that colour Namaqualand's veld if there has been sufficient rainfall there in winter. The spectacle in Kirstenbosch is an echo of Namaqualand's famous spring flowering but happens reliably every year. Note that the flowers are heliotropic, turning to follow the sun as it moves across the sky; they look their best in full sun from mid-morning to mid-afternoon.

While the most spectacular display is in the beds around the main lawns, there are brilliant patches of colour all over the unforested areas of the garden at this time of year.

As spring becomes summer and the annuals fade, large plantings of watsonias, perennial bulbous plants of the Iridaceae family, flower in white and shades of pink, orange and red. These in turn give way in full summer to the mass displays of *Agapanthus* species and cultivars, tuberous perennials of the lily family with flowers of white and many unusual shades of blue. Summer is also the season when the brilliant royal blue flowers of *Aristea major* appear.

A bank of white plectranthus invites you into the Camphor Avenue.

The forest lily (*Clivia miniata*).

The George (or Knysna) lily (*Cyrtanthus elatus*).

THE CAMPHOR AVENUE

The Camphor Avenue seldom fails to entrance visitors. It is permanently shaded by camphor trees (*Cinnamomum camphora*) planted by Cecil John Rhodes when he owned Kirstenbosch, what is now known as Cecilia Forest to the south, and the Groote Schuur Estate to the north. These trees are not indigenous (they come from China and Japan), but their massive spreading branches provide a beautiful backdrop to the shade-loving plants grown in raised beds beneath them. The avenue is at its most enchanting in March when massed plantings of *Plectranthus* species flower in waves of white, delicate lilacs and mauves – a spectacular sight in the subdued light. There is colour here for most of the year, however, with a rich variety of some of South Africa's most splendid shade-loving bulbous plants. Most of these belong to the Amaryllidaceae family and they flower in succession from late winter to autumn – the white 'paintbrushes' of *Haemanthus deformis* and *Haemanthus albiflos*, the dusky pink spikes of *Veltheimia bracteata* (a member of the lily family), the brilliant orange trumpets of the forest lily (*Clivia miniata*), the bright red flowers of *Scadoxus puniceus* (snake lily), *Scadoxus multiflorus* subsp. *katherinae* (blood flower) and *Scadoxus membranaceus* (paintbrush) with its large brown bracts, the pink trumpets of *Crinum* species and, in autumn, the scarlet flowers of the George lily (*Cyrtanthus elatus*). Shade-loving cycads are also a feature of these beds. At either end of the avenue are plantings of bulbs that enjoy less shade and provide a colourful entrance.

Scadoxus multiflorus subsp. *katherinae.*

Chasmanthe floribunda var. *duckittii* flower below Morton Bay figs (*Ficus macrophylla*).

Plectranthus and a cycad under the camphor trees.

Spring daisies in the lower garden.

THE LOWER GARDEN

The main lawns cover the area of Kirstenbosch which is most accessible from the main gate and parking area. This is a spacious part of the garden, dominated by large oaks, most of which were planted over a century ago. These are not indigenous to South Africa, but they add a special character to this area where Kirstenbosch's unique setting under Table Mountain can best be appreciated. The lawns, planted with buffalo grass (*Stenotaphrum secundatum*, a local species which grows naturally on the Peninsula and elsewhere), are particularly beautiful in winter when sorrel (*Oxalis* spp.) flowers, covering large areas in sheets of pink or white.

The main pond is planted with indigenous water lilies and surrounded by water-loving plants, including *Crinum campanulatum* which produces pink trumpets in summer. The pond is fed by the perennial flow from the spring in the Dell. The streams from Nursery Ravine and Skeleton Gorge meet in the centre of the lawns. Reduced to a trickle in summer, these flow strongly after heavy winter rains.

On the eastern boundary of the lawns is a row of large Morton Bay fig trees (*Ficus macrophylla* from Australia, planted a century ago by Cecil John Rhodes) and next to this a copse of Cape and Natal hardwood trees. Below the head office of the Botanical Society of South Africa are two unique features of Kirstenbosch. In a marshy area next to the main drive to the parking area is the vlei garden, a planting of species from the western and southern Cape adapted to growing naturally in seeps – wet but not stagnant areas. The beds to the south of this are devoted exclusively to plants from the Cape Peninsula.

The indigenous water lily (*Nymphaea nouchali*).

The oaks in autumn colours.

In the forest along the Stinkwood Trail.

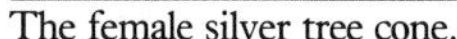

The female silver tree cone.

The male silver tree inflorescence.

THE FOREST TRAILS

The three forest trails are circular walks which have been laid out within the borders of Kirstenbosch. Each is well signposted. They require no hiking equipment except sensible clothes and walking shoes.

In the forest, specimen trees have been labelled with their common and botanical names.

THE STINKWOOD TRAIL
Length: 1,2 km
Time: 45 minutes (without stops)
Exertion: Mild
Start: At the top end of the parking area next to the Garden Shop

This short loop takes you past the Gold Fields Environmental Education Centre up into the natural forest on a steadily rising path, to a junction of four forest tracks. The return route follows one of these back to the Environmental Education Centre and the start.

THE YELLOWWOOD TRAIL
Length: 2,5 km
Time: 1,5 hours (without stops)
Exertion: Moderate
Start: At the north-west end of the parking area

Apart from small sections in fynbos, this short circuit leads you through natural forest to meet Smuts' Track which returns you to the main path next to the fragrance garden.

THE SILVER TREE TRAIL
Length: 6 km
Time: 3 hours (without stops)
Exertion: Moderate
Start: At the Rycroft Gate

This trail is aptly named as, along the first part past the protea garden on your right and the dam on the left, silver trees *(Leucadendron argenteum)* are abundant - as they are on the other parts of the route which pass through natural fynbos on the slopes of Table Mountain. After crossing Nursery Stream in natural forest, the route swings north, more or less on the contour, first in fynbos then in forest, crossing Skeleton and Window streams before dropping to a little-known and seldom visited part of Kirstenbosch - Lübbert's Gift. Here you take the Woodcutter's Path (a reminder of the days of the Dutch East India Company when wood was felled here) to meet the contour path. The most strenuous part of the trail starts at the junction, a zigzag climb to clear a scree where aloes grow among the sandstone boulders, and down again. From the crossing of Window Stream, the contour path keeps true to its name, and it is a virtually level walk to the southern boundary of Kirstenbosch, where a path leads you down past the dam and back to the start.

A female paradise flycatcher in the forest.

Erica empetrina grows along Smuts' Track, from where there is a view across the Back Table to Constantiaberg.

Erica curviflora flowers at the site of the old Lister Nursery above Nursery Ravine.

The contour path near Nursery Ravine.

Along the path from the Garden to Nursery Ravine, dominated by Castle Rock.

Kirstenbosch is one of the starting points for hikes along and up the eastern face of Table Mountain. There is a wide variety of routes from here to all parts of the mountain, including the cable station, front face, western side, and the Back Table and Constantia Nek. Brief descriptions of three representative walks which start and end at Kirstenbosch follow. Consult the map on page 4.

Aristea major flowers in a recently burnt patch of fynbos beside Smuts' Track.

CECILIA FOREST

Time: 3 hours, excluding stops
Exertion: Moderate
Start: The Rycroft Gate

From the Rycroft Gate, this route leads past the southern side of the dam to meet the contour path. Just to the right, a signposted path leads up off the contour path across a wooded kloof to Cecilia Ravine. At the next path junction, keep right up to the pines where a track leads into the plantation to the concreted Bridle Path. Follow this downwards, swinging sharp left down a track to meet another track, which here is the contour path. Turn sharp right and, at the next junction, sharp left past several cork oaks (part of Cecil Rhodes' old road), past the forestry station to the contour path, which leads you back to Kirstenbosch again.

THE RESERVOIRS

Time: 3,5 hours, excluding stops
Exertion: High
Start: The Rycroft Gate or the parking area

Follow the signs to the contour path and Nursery Ravine. Next to the rocky bed of the stream in Nursery Ravine the path starts its steep, zigzagging ascent – this is strenuous, but the only really hard part of the route. At the top of the ravine ignore the minor paths to the left and right, and walk through the former Lister Nursery site (an old proof site for testing the adaptability of exotic trees for the then planned afforestation of Table Mountain) to the concreted road which runs to the Hely-Hutchinson Reservoir on your right. Turn left along the road and follow this past the three smaller reservoirs and off the Back Table. In the Cecilia pine plantation keep sharp left at both junctions of forestry tracks. The second track ends at the northern edge of the trees and becomes a path on the contour. This soon meets a path to the right which leads steeply down to a track which takes you directly to the contour path. Turn left onto this path and continue back to the Kirstenbosch boundary, from where signposts will return you to the start.

SMUTS' TRACK

Time: 5,5 hours
Exertion: Very high
Start: The Rycroft Gate or the parking area

From the start, follow the signs to the contour path and Skeleton Gorge. Next to the stream bed in Skeleton Gorge an old sign 'Smuts' Track' marks the start of the steep climb up the gorge. A bit of easy rock-scrambling, with the help of strategically placed exposed tree roots, is required, and at one point a fixed ladder helps you up and over a large rock. Once out of the forest at the top of the gorge you have a choice. An easy route back to Kirstenbosch is the path which continues straight on to meet the fence around the Hely-Hutchinson Reservoir. This path meets the one from the old Lister Nursery and the top of Nursery Ravine (on the left), the latter providing an easy but steep way back to Kirstenbosch.

If you want to reach the highest point on Table Mountain, Maclear's Beacon, take one of the two paths leading north (right) at the top of Skeleton Gorge. These meet later, just before you cross the stream leading to Window Gorge (Window Gorge is dangerous!). Keep to the right on the main path which leads easily up and over two rock faces and continue along Smuts' Track, past the junction with a path on the left to Echo Valley, to Maclear's Beacon. To return to Kirstenbosch, retrace your steps and descend Skeleton Gorge or Nursery Ravine as described earlier.

The red disa, *Disa uniflora.*

THE NATIONAL BOTANICAL INSTITUTE

The National Botanical Institute (NBI) was established in 1989 as a result of the amalgamation of the National Botanical Gardens and the Botanical Research Institute. There are eight national Botanical Gardens - three in the Western Cape, one in the Free State, one in Kwa-Zulu Natal, two in Gauteng and one in Mpumalanga. The NBI also runs a network of herbaria, research units and education centres round the country.

The mission of the National Botanical Institute is to promote the sustainable use, conservation, appreciation and enjoyment of the exceptionally rich plant life of South Africa for the benefit of all people.

FACTS AND FIGURES

Kirstenbosch is located at 33° 59′ South, 18° 26′ East on the eastern slopes of Table Mountain.

Its total area is 530 ha, of which 478 ha are natural (forest or fynbos), 36 ha are cultivated and 16 ha are utilised for buildings, services and facilities.

The rocks of the cliffs are Table Mountain sandstone and the very acidic soils of the higher area are derived from these. The soils of the cultivated area are acid and either granitic in origin or shale-derived, the latter being the more fertile.

Kirstenbosch was the first major botanical garden in the world to be devoted to indigenous (local) plants and there are nearly 6 000 species of southern African plants growing in the cultivated area of the Garden, while some 900 species of plants flourish in the natural areas. Kirstenbosch is essentially a garden in which wild southern African plants are introduced to the public and assessed for their horticultural potential or other value. Plants of the Cape Flora are well represented, but in addition there are others from all over southern Africa which have adapted to Kirstenbosch's climate and soils.

The main gate to Kirstenbosch.

The Garden is clearly and comprehensively signposted.

THE BOTANICAL SOCIETY

The Botanical Society of South Africa plays a key role in support of the National Botanical Gardens. The head office is located just below the Garden Centre. The office of the Kirstenbosch branch of the society is located in one of the stone cottages below Rhodes Avenue (see map on page 2). Anyone interested in southern Africa's flora should consider joining the society. Benefits of membership include:

- ☐ Free entry to all the National Botanical Gardens
- ☐ 10% discount on plants, books and gifts
- ☐ Free attendance at all Botanical Society outings, lectures and other events.
- ☐ Receipt of *Veld & Flora*, the society's quarterly journal.
- ☐ Discount on indigenous plant seeds.

Scholars relax next to the Gold Fields Environmental Education Centre.

INFORMATION

The National Botanical Institute accepts no liability for any personal injury or accident to members of the public, or for loss or damage to their personal property in Kirstenbosch. Visitors should note that there are poisonous snakes and open areas of water in the Garden; hikers on Table Mountain should take the necessary care (see page 5).

ACCESS AND DIRECTIONS (see map page 32)

PUBLIC TRANSPORT

A bus service links the main gate with Mowbray railway station. For details, telephone the information desk (tel. 021-761 4916)

OPENING TIMES

September to March 08h00 to 19h00;
April to August: 08h00 to 18h00.
Late closing for summer concerts.

ENTRANCE FEE

There is a charge per person to enter the Garden. *Bona fide* pensioners have free entry on Tuesdays. Members of the Botanical Society of South Africa, Western Province Mountain Club and the Mountain Club of South Africa enjoy free entry, but must produce a valid membership card at the gate.

ADMISSION OF PETS

Dogs are not permitted in the Garden without a permit from the curator's office. Failure to comply may result in being fined.

SIGNPOSTING

Focal areas of the Garden, the three forest trails and features on the east face of Table Mountain are all well signposted.

FACILITIES

The information desks (tel.: 021-761 4916) are located at the Visitors' Centre and Garden Centre gates. Lost property should be reported, handed in and claimed here. This is also the first place to contact in an emergency.

Security guards, who are in contact with the information desk by radio, patrol the Garden.

There are public toilets at several points in the Garden (see map on page 2).

The Kirstenbosch Shop at the Visitors' Centre (tel. 021-762 2510) sells gifts, cards, guide books and a variety of books about plants and gardens. The Garden Centre (tel. 021-762 1621) sells indigenous plants and gardening accessories.

REGULATIONS

Kirstenbosch is managed as a nature reserve for visitors to enjoy in peace. For this reason:

- ☐ No games of any nature are allowed. Wheeled toys, e.g. roller skates and skateboards, are prohibited.
- ☐ Neither vehicles (including cycles) nor horses are allowed beyond the car park.
- ☐ The lighting of fires, picnicking (except during summer sunset concerts) and littering are prohibited.
- ☐ No radios, tape recorders or musical instruments are allowed in the Garden.
- ☐ Damage to plants (including flower-picking and seed harvesting) or animals is strictly prohibited.
- ☐ All weapons are strictly prohibited.
- ☐ Any behaviour or activity that disturbs other visitors is prohibited.

FOR THE DISABLED
For the convenience of the visually impaired and the elderly, the braille trail and the fragrance garden [8/9] have been established. It is easier, because of the steep slopes, to tour the Garden in a wheelchair from the Rycroft Gate. Wheelchairs are available at the Rycroft Gate and the other entrances but they should be booked in advance (tel. 021-761 4916). Entry from the Rycroft Gate is also recommended for the frail.

A small electric vehicle takes visitors on guided tours around the Garden very day of the week on request. tel.: (021) 761 4916.

GARDEN TOURS
Free guided garden walks start at 11h00 on Tuesdays and Saturdays. Enquire at the Information desks. Tel.: 021-761 4916.

RESTAURANT
The restaurant (tel.: 021-797 7614) serves, breakfasts, lunches, teas and snacks seven days a week (closed Christmas Day). Caffé Botanica (tel.: 021-7626841) serves light snacks and drinks.

ACTIVITIES
Weddings, conferences and other functions may be held in the Old Mutual Conference Centre. Write to the curator, Kirstenbosch National Botanical Garden. A charge is levied for this facility.

The Sanlam Lecture Hall near the Garden Centre entrance can be hired for functions and shows. Contact the curator's office (tel 021-762 9120).

Indigenous flower shows are held annually at certain seasons, as are shows of exotic flowers by various horticultural societies.

The area below Rhodes Avenue (see map on page 2) is the venue for a regular craft market and other similar activities. Contact the information office for details.

Summer sunset concerts (December to March) feature a wide variety of music ranging from classical through African to jazz. Visitors are invited to picnic on the lawns during the concerts on Sunday evenings.

The nursery, Kirstenbosch.

The Gold Fields Environmental Education Centre conducts numerous programmes throughout the year. Details of these and the summer concerts may be obtained by telephoning 021 762 1166.

ADDRESSES
The Kirstenbosch National Botanical Garden and the National Botanical Institute share the same address:
Private Bag X7, Claremont, 7735

Contact numbers:
Kirstenbosch
Tel. (021) 762-9120; Fax (021) 797-6570
National Botanical Institute
Tel. (021) 762-1166; Fax (021) 762-3229

The Botanical Society of South Africa:
Private Bag X10,
Newlands, 7735
Head Office: Tel. (021) 797 2090; Fax (021) 797 2376
Kirstenbosch branch: Tel. (021) 615 468

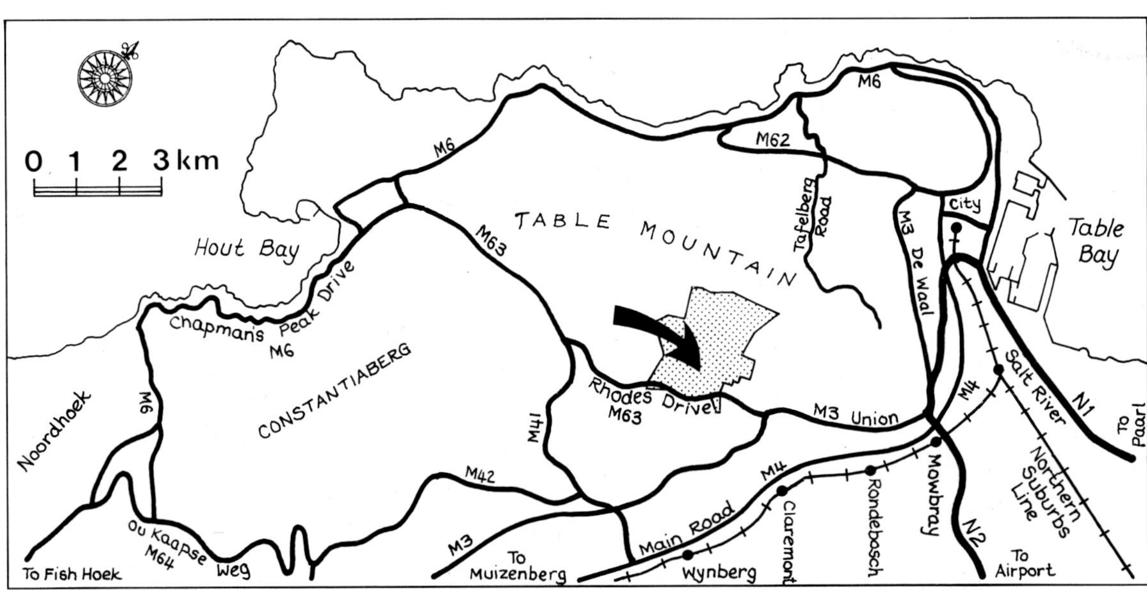

ACCESS MAP

Published by The National Botanical Institute, Private Bag X7, Claremont 7735, South Africa.

Concept by Colin Paterson-Jones.
Designed by Linda Vicquery. Edited by Tessa Kennedy.
Reproduction by Unifoto, Cape Town.

ISBN: 1-874907-13-7